Extreme Temperatures

Learning About Positive and Negative Numbers

John Strazzabosco

Math
for the
REAL World™

Rosen Classroom Books & Materials
New York

Published in 2004 by The Rosen Publishing Group, Inc.
29 East 21st Street, New York, NY 10010

Book Design: Michael Tsanis

Photo Credits: Cover, pp. 18, 30 (Antarctica) © John Giustina/Taxi; p. 4 © Pete Turner/The Image Bank;
p. 10 © Wolfgang Kaehler/Corbis; p. 12 © Marc Garanger/Corbis; p. 16 © Art Wolfe/The Image Bank; p. 20
© Galen Rowell/Corbis; p. 22 © Digital Vision; p. 24 (penguins) © Peter Scoones/Taxi; p. 24 (seal) ©
Cousteau Society/The Image Bank; p. 24 (whales) © Corbis; p. 26 (International Space Station) © AFP/
Corbis; pp. 26 (Earth), 28 © PhotoDisc; p. 30 (Sahara desert) © Andrea Pistolesi/The Image Bank.

Library of Congress Cataloging-in-Publication Data

Strazzabosco, John.
  Extreme temperatures : learning about positive and negative numbers /
John Strazzabosco.
        v. cm. — (PowerMath)
Includes index.
Contents: How low can it go? — The Arctic — A cold desert — The world's
greatest chill — Out of this world — From one extreme to another.
  ISBN 0-8239-8997-6 (lib. bdg.)
  ISBN 0-8239-8927-5 (pbk.)
  ISBN 0-8239-7455-3 (6-pack)
  1. Numbers, Real—Juvenile literature. 2. Low temperatures—Juvenile
literature. [1. Numbers, Real. 2. Low temperatures.] I. Title. II.
Series.
  QA141.S84 2004
  512'.7—dc21
                              2003001768

Manufactured in the United States of America

Contents

Thermometers can measure the temperature inside as well as the temperature outside. Do you have a thermometer in your home?

°F

100
90
80
70
60
50
40
30
20
10
0
−10
−20
−30
−40

Let's get a feel for measuring degrees with the Fahrenheit scale by looking at some Fahrenheit temperature facts:

- **32°F is the temperature at which water freezes.**
- **80°F is just right for swimming at the beach.**
- **212°F is the temperature at which water boils.**

How Low Can It Go?

To find extreme cold temperatures, we can travel to extreme places. We can go to the island of Greenland in the North Atlantic Ocean, to a region in northern Asia called Siberia, or to northern Alaska in the United States. We can go to a desert in Asia called the Gobi desert, or to the coldest place on Earth—Antarctica. We might even travel to outer space!

An instrument called a **thermometer** measures temperature. Temperature is measured in **units** called **degrees**. Thermometers in the United States generally measure degrees using what is called the Fahrenheit scale. In most other countries, the Celsius scale is used. When we measure degrees, we attach a special **symbol** after the number of degrees. The symbol looks like this: °

When we measure degrees using the Celsius scale, we attach °C after the number of degrees. When we measure degrees using the Fahrenheit scale, we attach °F after the number of degrees. In this book we'll use the Fahrenheit scale.

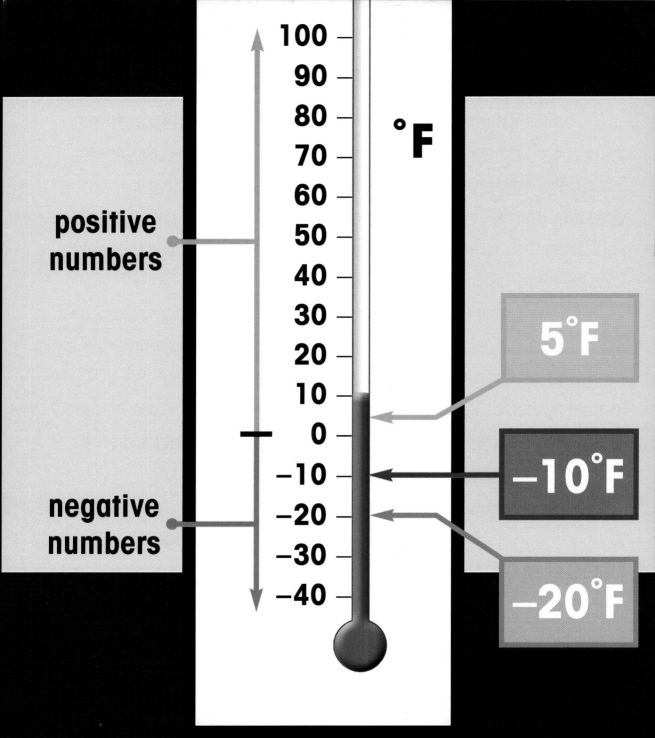

positive
numbers

negative
numbers

°F

100
90
80
70
60
50
40
30
20
10
0
−10
−20
−30
−40

5°F

−10°F

−20°F

A thermometer measures degrees down to zero, but it doesn't stop there. If the weatherman says that it is "10 below" on some cold winter night, he means that it is 10°F below zero. That temperature measurement is often expressed as −10°F, which we read as "minus 10 degrees Fahrenheit." Numbers above zero are called positive numbers. Numbers below zero are called **negative** numbers.

Negative numbers are written with a negative symbol in front of them. Sometimes a negative symbol is called a minus sign. You have seen this sign in math class when you do subtraction problems. Let's look at the temperature −20°F to see how this works. This temperature could be called "minus 20 degrees Fahrenheit," "negative 20 degrees Fahrenheit," or simply "20 below." All of these mean the same thing.

The red box at the left points to a temperature of −10°F. The blue box shows the temperature if we drop down another 10°F to −20°F. What is the temperature if we go up 15°F from −10°F? You can look at the green box to find the answer.

Greenland's inner plateau would have to drop 34°F to go from its average February temperature of –53°F to its coldest recorded temperature, –87°F.

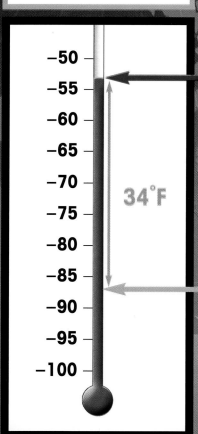

−50
−55 ← −53°F
−60
−65
−70
−75   34°F
−80
−85
−90 ← −87°F
−95
−100

Cape Morris Jesup

Arctic Oce

Greenland

The layer of ice covering Greenland's inner plateau has an average thickness of 1 mile. In some places it is up to 2 miles thick!

The Arctic

To get a better idea about how negative temperatures work, we can take a look at some of the world's coldest places, like the island of Greenland that lies between the North Atlantic Ocean and the Arctic Ocean. Greenland is the largest island in the world, and is very close to the North Pole. The North Pole is located in the northernmost part of Earth's Northern **Hemisphere**, in the Arctic Ocean. Cape Morris Jesup in Greenland is only about 440 miles from the North Pole. It is the most northern land in the world!

The name "Greenland" might make you think that the temperatures there would be warm and the land would be green and covered with plants, but this is not true. A low **plateau** covers about 80% of the island and is surrounded by mountains along the island's coastline. The inner plateau is always covered by a layer of ice, and very little can grow there. Greenland's coldest spot is found in the center of the ice-covered plateau. There, temperatures average about –53°F in February and about 12°F in July. The coldest temperature ever recorded in Greenland was –87°F, which was recorded in 1954.

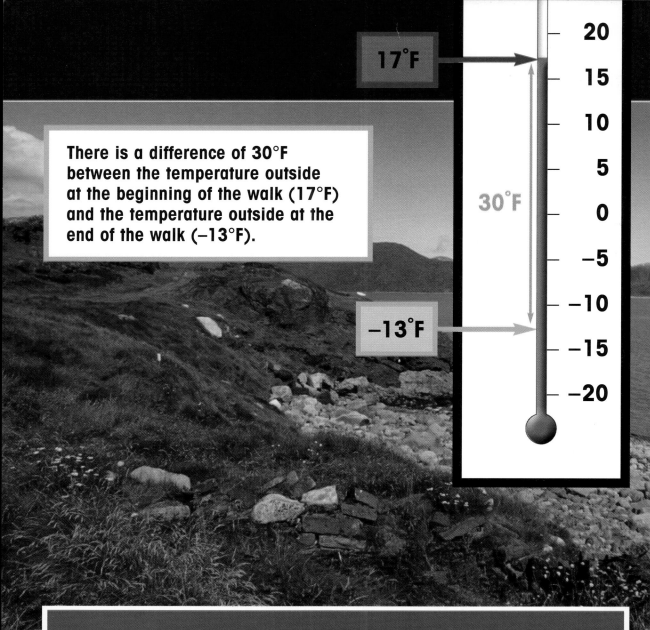

There is a difference of 30°F between the temperature outside at the beginning of the walk (17°F) and the temperature outside at the end of the walk (−13°F).

17°F

30°F

−13°F

Greenland has a very short summer. Despite the island's name, only the areas along the coasts are actually green.

Greenland isn't always cold. The short summer lasts from mid-July to early September. During summer, Greenland's temperatures average about 50°F on the southwestern coast. This is where most of Greenland's people live, since it is the warmest part of the island.

Greenland has a population of about 60,000 people. This population is largely made up of a group of people called the Inuit (IH-nyoo-wuht). Temperatures can get as low as −50°F, so people cannot rush as they work. They would sweat beneath the fur clothing they must wear to keep warm. When they slow down, this sweat would freeze on their skin. This makes outside activity in Greenland difficult and even dangerous during the coldest parts of the year.

Let's say someone in Greenland decides to take a walk outside when the temperature is 17°F. The temperature drops to −13°F by the time the person finishes their walk. How many degrees has the temperature dropped altogether? We can add the number of degrees above zero (17°F) and the number of degrees below zero (13°F) to find out.

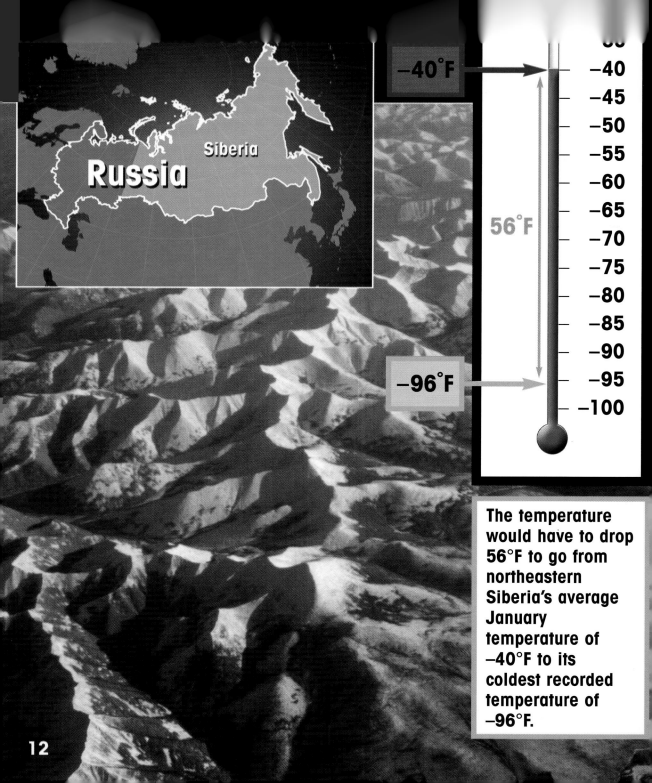

Russia

Siberia

−40°F

56°F

−40
−45
−50
−55
−60
−65
−70
−75
−80
−85
−90
−95
−100

−96°F

The temperature would have to drop 56°F to go from northeastern Siberia's average January temperature of −40°F to its coldest recorded temperature of −96°F.

Another of the Arctic's coldest places is a vast region in northern Asia called Siberia. Siberia covers about 75% of Russia, but not many people live there because it is so cold. For half of the year, much of Siberia is covered by snow and ice. The upper region of Siberia might experience a temperature of −65°F in winter, and the ground can remain frozen year-round.

The coldest part of Siberia is in the northeast. The average temperature there in January is −40°F! Temperatures there can range from about −59°F in January to 59°F in July. A low temperature of −96°F has been recorded there, which is colder than any temperature ever recorded at the North Pole!

In southwest Siberia, the winters are very cold but the summers can be very warm. Temperatures can be as cold as −40°F in January and more than 90°F in July!

An area of Siberia called the East Siberian Uplands is made up of a series of mountain ranges. Siberia's highest point can be found there—an active volcano that is 15,584 feet high!

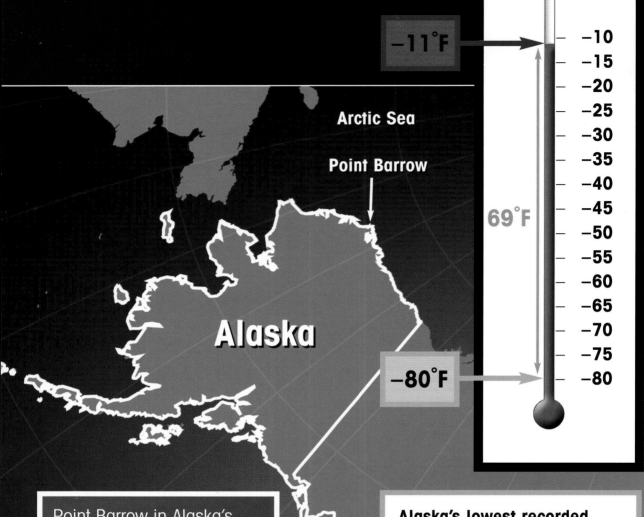

-11°F

Arctic Sea

Point Barrow

69°F

-80°F

Alaska

-10
-15
-20
-25
-30
-35
-40
-45
-50
-55
-60
-65
-70
-75
-80

Point Barrow in Alaska's Arctic region is the northernmost point in the United States. During the summer, this region has daylight for 24 hours a day!

Alaska's lowest recorded temperature of −80°F was 69°F colder than the Alaskan Arctic's average January temperature of −11°F.

Even the United States has places with extreme temperatures, like the state of Alaska. Alaska is the largest and northernmost state in the United States, and it covers an area that is almost $\frac{1}{5}$ as large as the rest of the country. Almost $\frac{1}{3}$ of Alaska lies in the Arctic region. As in Siberia, much of this land is frozen year-round. In the summer, the land's surface **thaws**. This allows wildflowers, grasses, and small shrubs to grow. However, trees cannot grow there. Trees need to put roots deep into the soil, and the ground below the surface is always frozen.

Alaska's Arctic region has an average temperature of −11°F in January. Alaska's northernmost spot is a place called Point Barrow. Although Point Barrow is almost 1,300 miles south of the North Pole, it still experiences some of the world's coldest temperatures. Winter temperatures there sometimes get as cold as −50°F. On January 23, 1971, a low temperature of −80°F was recorded close to Point Barrow in a place called Prospect Creek.

Mongolia

Gobi desert

China

A number line is a lot like a thermometer. It can show us the difference between positive and negative numbers.

A Cold Desert

You may not think of a desert when you think about cold places. We usually think of a desert as a hot, dry place, but a desert can also be a cold, dry place. The Gobi desert in central Asia, for example, has some of the lowest winter temperatures on Earth. The Gobi covers an area of more than 500,000 square miles, stretching across the southern part of Mongolia and the northern part of China. The middle part of the Gobi is mostly covered by dry, rocky, or sandy soil and is surrounded by dry, treeless plains. Most of the desert's plant and animal life can be found in the plains, which receive more rainfall than the desert's dry middle region.

In the Gobi, the range of temperatures throughout the year is extreme. In July, the temperature can reach 122°F. In January, the low temperature can reach −40°F. That is quite a difference!

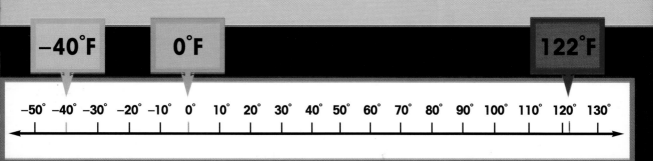

-40°F   0°F   122°F

-50° -40° -30° -20° -10° 0° 10° 20° 30° 40° 50° 60° 70° 80° 90° 100° 110° 120° 130°

Antarctica is surrounded by water and giant icebergs. The continent itself is mostly covered by huge glaciers and snow.

The World's Greatest Chill

We've taken a look at a few of the coldest places on Earth, but there is a place that is colder than all of them—the ice-covered **continent** of Antarctica, the land that surrounds the South Pole in the southernmost part of Earth's Southern Hemisphere. Antarctica is even colder than the North Pole. It is the coldest, windiest, driest land on Earth.

Antarctica has an area of about 5,400,000 square miles, which makes it larger than either of the continents of Australia or Europe. Most of Antarctica's land area is made up of a layer of ice that averages more than 7,000 feet thick. That is more than 1 mile! This layer of ice adds greatly to Antarctica's land area. Without it, Antarctica would be the smallest continent on Earth. This icy layer also gives Antarctica the highest elevation of all the continents— about 7,500 feet above sea level. About 98% of Antarctica is covered by ice and snow. Underneath the ice and snow lie mountains, valleys, and plains.

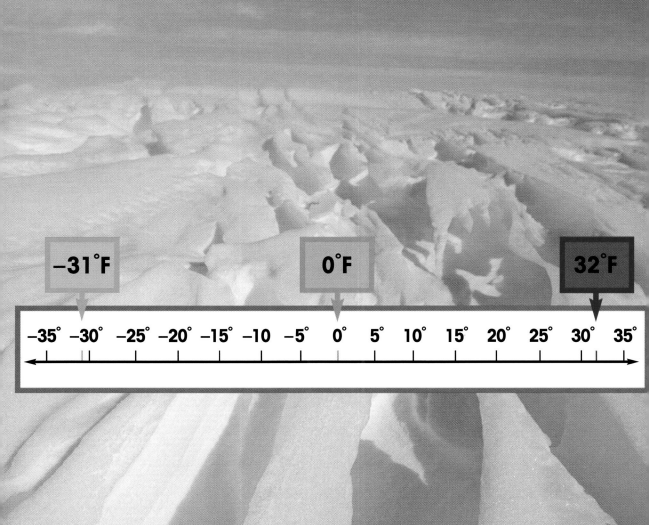

−31°F   0°F   32°F

−35° −30° −25° −20° −15° −10 −5° 0° 5° 10° 15° 20° 25° 30° 35°

The layer of ice that covers most of Antarctica has formed over millions of years as layers of snow were pressed together. In some areas, this ice is over 15,000 feet thick. That is almost 3 miles! Antarctica's icy layer contains 70% of the world's fresh water. If that ice melted, its water would drain into the world's oceans, making ocean levels rise by about 220 feet. That is about the same height as a 22-story building! Many cities and towns on coasts around the world would be completely flooded.

Antarctica's inner plateau is extremely cold and dry, and gets only about 2 inches of snow each year. Antarctica's coastline has milder conditions, with slightly warmer temperatures and about 24 inches of snowfall each year.

Even Antarctica's warmest summer temperatures are still cold, rarely reaching above freezing, or 32°F. The continent's summer lasts from December through February. During the summer, temperatures along the coast may reach 32°F, but inland, the temperature can be −31°F.

Let's look at the number line on page 20 to see the difference between −31°F and 32°F.

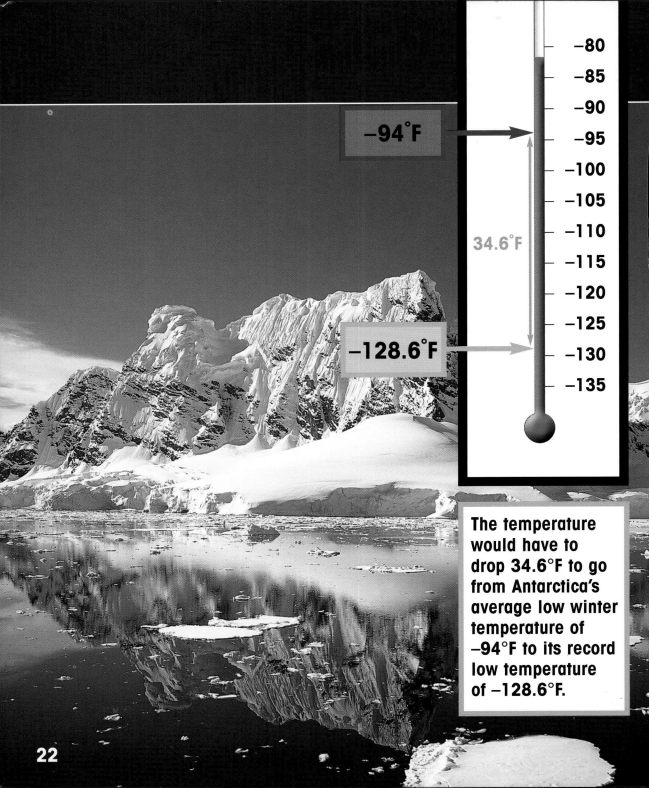

−94°F

34.6°F

−128.6°F

-80
-85
-90
-95
-100
-105
-110
-115
-120
-125
-130
-135

The temperature would have to drop 34.6°F to go from Antarctica's average low winter temperature of −94°F to its record low temperature of −128.6°F.

Antarctica's winter lasts from May through August and is the coldest part of the continent's year. During the winter, Antarctica's coastline temperatures usually range between −5°F and −22°F, while the inland plateau experiences temperatures between −40°F and −94°F. Bitter cold winds make these temperatures feel even colder. Winds can reach a speed of over 100 miles per hour as they blow down from the inland plateau toward the coast!

The coldest temperature ever recorded on Earth was measured in Antarctica on July 21, 1983. That temperature was −128.6°F! In contrast, some of the islands around Antarctica have recorded a high temperature of 50°F.

Because Antarctica is in the Southern Hemisphere, its seasons are the opposite of the seasons in the Northern Hemisphere. When it is wintertime in the Southern Hemisphere, it is summertime in the Northern Hemisphere.

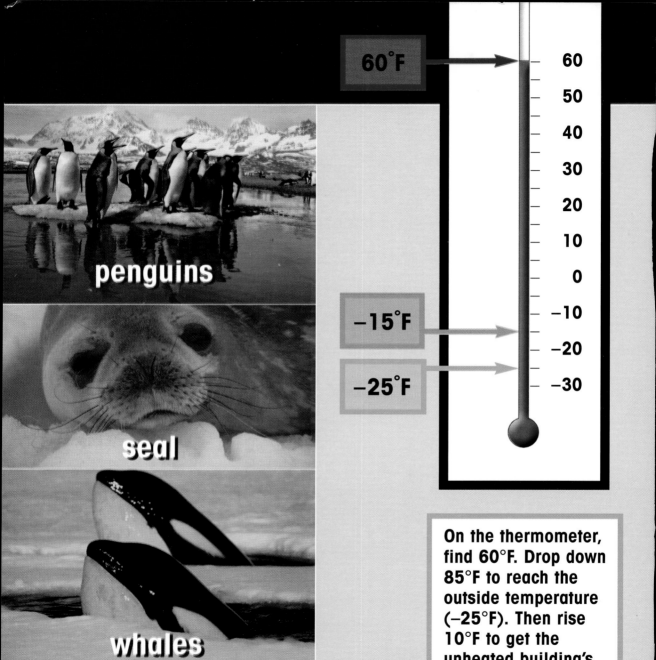

60°F

60
50
40
30
20
10
0
−10
−20
−30

−15°F

−25°F

penguins

seal

whales

On the thermometer, find 60°F. Drop down 85°F to reach the outside temperature (−25°F). Then rise 10°F to get the unheated building's temperature (−15°F).

The waters around Antarctica are home to whales, penguins, seals, and different kinds of fish. Tiny, shrimplike animals called **krill** also live in Antarctica's waters. Krill provide a source of food for the other animals and birds that live in the region.

One kind of fish called a toothfish lives off the Antarctic shore. It can grow to be 6 feet long and can live up to 2 miles below the ocean's surface at water temperatures that are near freezing. You might think this fish would freeze into a solid block of ice at such cold temperatures, but its blood contains a special **protein** that keeps it flowing.

The only people who live in Antarctica are scientists from around the world who study the region's plants, animals, and weather. Let's say a scientist's living quarters have a temperature of 60°F. The scientist steps outside to an 85°F drop in temperature, then enters an unheated building that is only 10°F warmer than it is outside. Find the outside temperature and also the second building's temperature.

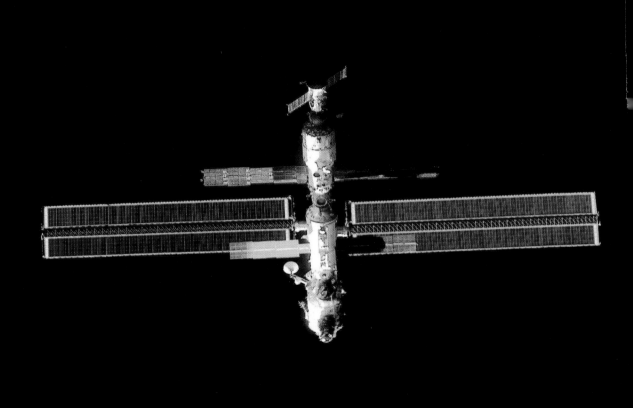

Out of This World

For really extreme temperatures, we can travel an extreme distance to the International Space Station, which is located about 240 miles above Earth. Many different countries are helping to build the International Space Station. When it is completed, astronauts and scientists will be able to live in the space station and do experiments there.

The station was specially designed to handle temperatures within the range of −250°F to 250°F. The station's fuel is stored at a temperature of −423°F, but the temperature of the fuel rises to 6,000°F when it burns! Therefore, the station's equipment must perform well under extreme hot and cold conditions to keep the people aboard safe.

When the International Space Station is finished, 6 people at a time will be able to live and work there.

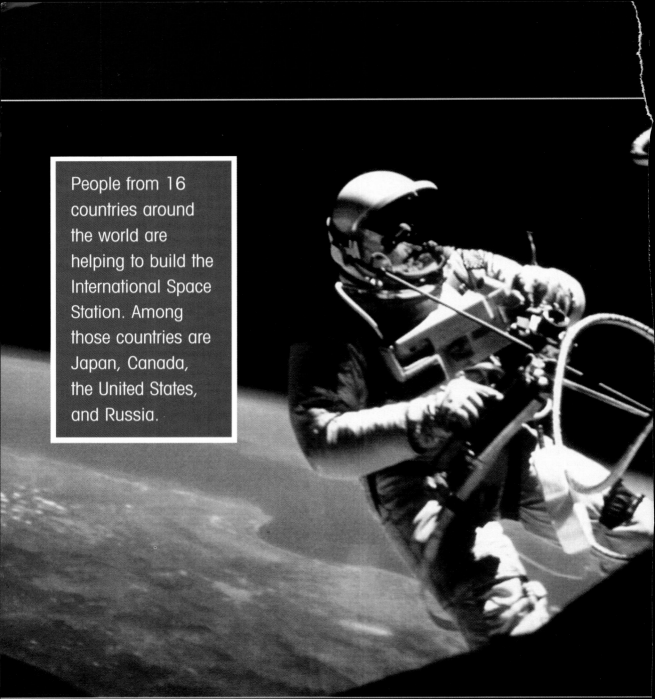

People from 16 countries around the world are helping to build the International Space Station. Among those countries are Japan, Canada, the United States, and Russia.

Because the International Space Station is being built in outer space, sometimes the people working on it must do work on the outside of the station. They must wear special spacesuits to protect their bodies from the extreme temperatures of space. Strong direct sunlight can make the outside of the spacesuit heat up to 300°F. If a person is working outside the space station but is not in direct sunlight, the outside temperature of the spacesuit could drop down to −200°F.

Let's look at a number line to get a better idea of the difference between these numbers.

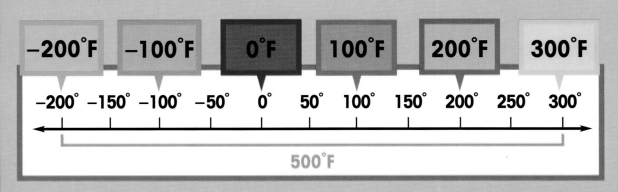

That's a range of 500°F! The spacesuit controls the temperature inside of it so the person stays safe and comfortable.

Back on Earth, we can compare the coldest temperature ever recorded on the planet in Antarctica (−128.6°F) with the hottest temperature ever recorded on the planet, which happened in the Sahara desert in Africa on September 13, 1922. That temperature was 136°F!

Now let's compare the United States' lowest recorded temperature of −80°F, which happened near Point Barrow, Alaska, with its hottest recorded temperature of 134°F, which happened in Death Valley, California, on July 10, 1913.

Those are some big differences in temperature. What is the lowest recorded temperature and the highest recorded temperature in the place where you live? You can look on a thermometer to figure out the difference between them!

136°F   −128.6°F

Glossary

continent (KAHN-tuhn-uhnt) One of the 7 large landmasses on Earth.

degree (dih-GREE) A unit for measuring temperature.

hemisphere (HEH-muh-sfeer) One half of Earth. Sometimes people talk about Earth's Northern and Southern Hemispheres. Other times they talk about Earth's Eastern and Western Hemispheres.

krill (KRILL) Tiny, shrimplike animals that live in the ocean.

negative (NEH-guh-tiv) Less than zero.

plateau (pla-TOH) A large, flat area of land in the mountains or high above sea level.

protein (PROH-teen) One of the elements that occur naturally in all animals and are necessary for the animal's health.

symbol (SIM-buhl) Something that stands for something else.

thaw (THAW) To melt.

thermometer (thur-MAH-muh-tur) A tool that measures how hot or cold something is.

unit (YOO-nit) A standard amount by which things are measured.